Canines in the Wild

JACKALS IN THE WILD

Natalie Humphrey

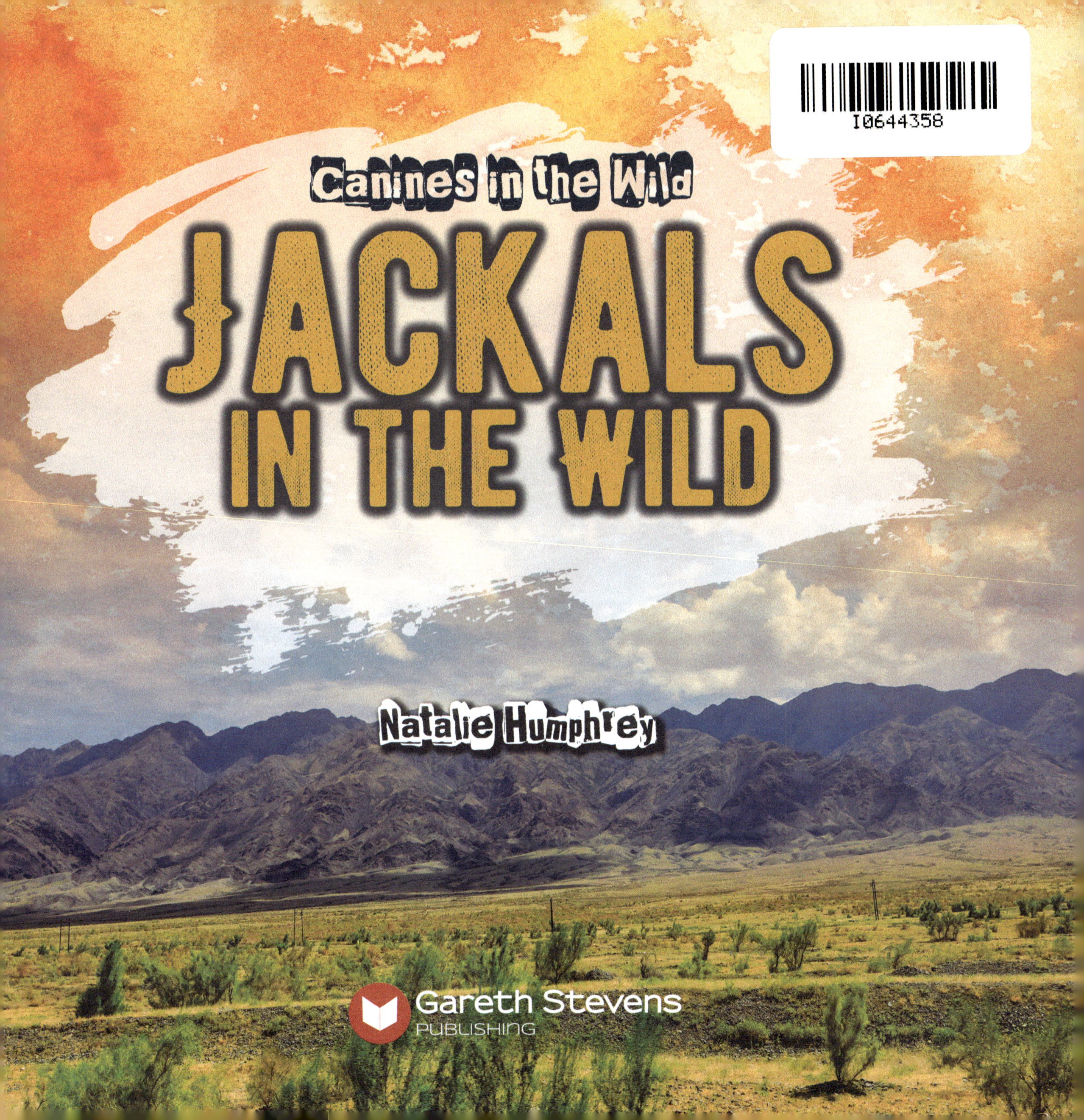

Gareth Stevens
PUBLISHING

Please visit our website, www.garethstevens.com. For a free color catalog of all our high-quality books, call toll free 1-800-542-2595 or fax 1-877-542-2596.

Cataloging-in-Publication Data
Names: Humphrey, Natalie.
Title: Jackals in the wild / Natalie Humphrey.
Description: New York : Gareth Stevens Publishing, 2023. | Series: Canines in the wild | Includes glossary and index.
Identifiers: ISBN 9781538279670 (pbk.) | ISBN 9781538279694 (library bound) | ISBN 9781538279687 (6pack) | ISBN 9781538279700 (ebook)
Subjects: LCSH: Jackals–Juvenile literature.
Classification: LCC QL737.C2 H86 2023 | DDC 599.77'2–dc23

Portions of this work were originally authored by David Lee and published as *Jackals*. All new material this edition authored by Natalie Humphrey.

Published in 2023 by
Gareth Stevens Publishing
29 East 21st Street
New York, NY 10010

Editor: Natalie Humphrey
Designer: Rachel Rising

Photo credits: Cover Zuzana Gabrielova/Shutterstock.com; cover, p. 1 ABCDstock/Shutterstock.com; cover, pp. 1, 3, 4, 6, 8, 10, 12, 14, 16, 18, 20, 22–24 Attitude/Shutterstock.com; cover, pp. 1, 3, 4, 6, 8, 10, 12, 14, 16, 18, 20, 22–24 donatas1205/Shutterstock.com; cover, pp. 1, 3, 6, 10, 12, 16, 18, 20, 22–24 Olga_C/Shutterstock.com; p. 4 Martin Mecnarowski/Shutterstock.com; p. 5 MZPHOTO.CZ/Shutterstock.com; p. 7 WildMedia/Shutterstock.com; p. 8 Rich Carey/Shutterstock.com; p. 9 Tomas Hejlek/Shutterstock.com; p. 11 Ondrej Prosicky/Shutterstock.com; p. 13 Yury Zap/Shutterstock.com; p. 14 Happy monkey/Shutterstock.com; p. 15 Madelein Wolfaardt/Shutterstock.com; p. 17 Ondrej Prosicky/Shutterstock.com; p. 19 Dmussman/Shutterstock.com; p. 21 Martin Prochazkacz/Shutterstock.com.

Printed in the United States of America

CPSIA compliance information: Batch #CSGS23: For further information, contact Gareth Stevens, New York, New, York at 1-800-542-2595.

CONTENTS

Words in the glossary appear in **bold** type the first time they are used in the text.

TINY TERROR

The jackal might not be the biggest wild **canine**, but this predator is a master of survival. Roaming the deserts and grasslands of the Eastern **Hemisphere**, the jackal is excellent at both staying out of sight and running when it needs to.

With a talent for finding food nearly anywhere, the jackal is known for being clever and sneaky. Like the fox or the coyote, the jackal is often shown as a trickster in stories. From India to Africa, the jackal is one wild canine!

The African jackal is called *mbweha* in Swahili, a language spoken in eastern Africa.

THREE DIFFERENT JACKALS

There are three different **species** of jackals living in different **habitats**. As long as there is a place for them to hide, jackals will live and grow! Black-backed jackals are found in **savannas** and woodlands in southern Africa. Side-striped jackals would rather live in wetter places, such as **marshes** in central and southern Africa.

Golden jackals, also called common jackals, are the only species found on more than one continent. They are usually found in the open grassy plains and deserts of South and Southeast Asia, and Europe.

That's Wild!

Until 2015, there was another species of jackal called the African golden jackal. After studying its **DNA**, scientists found it was a wolf, not a jackal at all!

The golden jackal's habitat is currently spreading into central and western Europe.

WHAT MAKES A JACKAL?

The jackal is one of the smallest canines in the world! Jackals have thin bodies and big, pointed ears. Most have bushy tails. They usually weigh between 17 and 25 pounds (8 and 11 kg). They're usually 16 inches (40 cm) tall and between 27 and 33 inches (70 and 84 cm) long.

Golden jackals have sand colored fur with black. The side-striped jackal has stripes along the sides of its body and a white tip on its tail. Black-backed jackals are sand-colored with a black back.

Jackals live between 8 and 9 years in the wild, but they can live up to 16 years in **captivity**.

A STRONG FAMILY

Like many canines, jackals are often social animals. While some jackals like to live and hunt alone, many others do so in pairs. Most jackals live and hunt in packs. These packs are strong family units that help keep the members safe and well fed.

Packs usually include a male, female, and their pups, or babies. Jackal couples often mate for life, which means they make babies only with each other. The pack sometimes includes older jackal babies that help care for their younger brothers and sisters.

That's Wild!

Like many wild dogs, jackals can tell the difference between the howls from their family and the howls from other jackals.

A jackal pack is usually around six jackals.

WHAT'S THAT SMELL?

When jackals are hunting alone, they may ambush their **prey**, or attack it by surprise. To do this, they roll in the dirt to cover their own scent, or smell. Then, they sneak up behind their prey and strike! Jackals grab their prey by the neck and shake it to kill it.

When hunting in a pack, jackals will howl to let their pack know when prey is nearby. When the pack is hunting together, it can take down much larger prey.

That's Wild!

Like most wild dogs, jackals keep and protect a territory with their pack. If another jackal or predator enters their territory, jackals will attack!

Jackals will often attack animals when the animals are sleeping.

AMAZINGLY ADAPTABLE

Jackals have **adapted** perfectly to their habitats. They have an amazing sense of sight, even in the dark. This helps jackals hunt at nighttime. They also have a great sense of hearing, thanks to their big ears. A jackal's sense of smell is possibly 100 times stronger than a human's!

A jackal's body is well built for roaming long distances, chasing prey, and running from bigger predators. Its fur color helps it **camouflage** in the dusty desert and dry grasslands.

Jackals are generally nocturnal, or most active at night.

EATING ANYTHING

Jackals aren't very picky about their meals. Instead, a hungry jackal will eat nearly anything it can find. A jackal hunting on its own might eat small animals, such as bugs, birds, reptiles, and small mammals. A jackal pack will hunt antelopes and gazelles. Jackals will also hunt and eat livestock, or farm animals.

When jackals aren't hunting animals, they eat fruits and grasses. They're also scavengers, which means they will eat dead animals left behind by larger predators. They'll even eat trash left behind by people!

That's Wild!

When a jackal hears or smells a successful hunt, it will often wait around the larger predator. Then, when the larger predator is not paying attention, the jackal will rush in and steal some meat!

Jackals can run up to 40 miles (64 km) per hour.

PUPS ON THE MOVE

Jackal parents usually raise their young together. A mother jackal gives birth to a **litter** of two to four pups. Jackals give birth in dens, which are usually holes in the ground dug out by the parents. Pups are born blind, and they are unable to care for themselves when they're born.

When the pups are around six months old, the jackal parents will teach them how to hunt. When a pup is close to one year old, it may leave the pack to start its own.

That's Wild!

Every two weeks, the jackal parents will leave their den and build a new one. The pups will be regularly moved like this until they're old enough to keep themselves safe.

Jackal pups weigh between 7 and 8 ounces (200 and 225 g) at birth.

LIFE ALONGSIDE JACKALS

While jackals are great hunters, they're also prey for larger predators. Hyenas, leopards, and eagles will hunt and eat young jackals when they're alone. Packs of jackals are usually safer.

One of the biggest dangers to jackals, though, is humans! Over the years, humans have moved farther and farther into land where jackals live. When jackals can't find enough food, they might wander onto a farm for food. Farmers will often poison or shoot these jackals to protect their livestock.

That's Wild!

Some organizations, like the African Wildlife Foundation, offer to help farmers live with jackals nearby. These organizations help build fences for livestock and help protect the land where jackals live.

Many people hunt jackals for their fur.

GLOSSARY

adapt: To change easily to suit new conditions.

camouflage: To blend in with one's surroundings.

canine: Of or relating to dogs.

captivity: The state of being caged.

DNA: Part of the body that carries genetic information, which gives the instructions for life.

habitat: The natural place where an animal or plant lives.

hemisphere: One-half of Earth.

litter: A group of pups born at the same time.

marsh: An area of soft, wet land.

prey: An animal that is hunted by other animals for food.

savanna: A grassland with scattered patches of trees.

species: A group of plants or animals that are all of the same kind.

FOR MORE INFORMATION

BOOKS

Castelló, José R. *Canids of the World: Wolves, Wild Dogs, Foxes, Jackals, Coyotes, and Their Relatives*. Princeton, NJ: Princeton University Press, 2018.

Stewart, Melissa. *Ick! Delightfully Disgusting Animal Dinners, Dwellings, and Defenses*. Washington, DC: National Geographic Kids, 2020.

WEBSITES

African Wildlife Foundation: Jackal
www.awf.org/wildlife-conservation/jackal
Learn more about jackals and how to protect them in the wild!

A to Z Animals: Jackal
www.a-z-animals.com/animals/jackal
Check out more interesting jackal facts.

Publisher's note to educators and parents: Our editors have carefully reviewed these websites to ensure that they are suitable for students. Many websites change frequently, however, and we cannot guarantee that a site's future contents will continue to meet our high standards of quality and educational value. Be advised that students should be closely supervised whenever they access the internet.

INDEX